Bioenergy Sustainability Conference 2019

AA008975

Nashville, Tennessee, USA
21 – 22 October 2019

ISBN: 978-1-7138-0568-7

Printed from e-media with permission by:

Curran Associates, Inc.
57 Morehouse Lane
Red Hook, NY 12571

Some format issues inherent in the e-media version may also appear in this print version.

Copyright© (2019) by AIChE
All rights reserved.

Printed with permission by Curran Associates, Inc. (2020)

For permission requests, please contact AIChE
at the address below.

AIChE
120 Wall Street, FL 23
New York, NY 10005-4020

Phone: (800) 242-4363
Fax: (203) 775-5177

www.aiche.org

Additional copies of this publication are available from:

Curran Associates, Inc.
57 Morehouse Lane
Red Hook, NY 12571 USA
Phone: 845-758-0400
Fax: 845-758-2633
Email: curran@proceedings.com
Web: www.proceedings.com

TABLE OF CONTENTS

ADVANCES IN INTEGRATED ASSESSMENTS OF BIOENERGY SUSTAINABILITY

INTEGRATING SYSTEMS DESIGNING SUSTAINABLE BIOENERGY SYSTEMS

KEYNOTE TALKS

RESEARCH FRONTIERS FOR SUSTAINABLE BIOENERGY SYSTEMS

POSTER SESSION

An Integrated Life Cycle Assessment Modeling Framework to Advance Biofuels and Bioproducts Production

BIOENERGY SUSTAINABILITY CONFERENCE

Conference: Bioenergy Sustainability Conference

Year: 2019

Proceeding: 2019 Bioenergy Sustainability Conference

Group: Abstract Submissions

Session: Advances in integrated assessments of bioenergy sustainability

Authors:

Shi, R., University of Illinois at Urbana-Champaign
Guest, J., University of Illinois at Urbana-Champaign

Life cycle assessment (LCA) has been widely used to evaluate the potential environmental impact associated with biofuels and bioproducts production. Traditionally, LCA studies adopt a static approach and apply rigid factors throughout the life cycle inventory (LCI) and life cycle impact assessment (LCIA) stages. This practice, however, often overlooks large uncertainty, particularly in the effect of multiple scales of system, various process parameters, and data sources, and significantly limits the generalizability of findings and the transferability of the results to other LCAs. In this study, we propose a new quantitative LCA approach by utilizing an open-source biorefinery modeling platform established in Python, which integrates biorefinery simulation, design basis algorithms, and LCA to elucidate optimal pathways for the production of biofuels and bioproducts. To demonstrate, a case study for sugarcane ethanol production will be presented. The integrated model has the ability to size and predict performances of each unit process based on detailed design decisions and operating characteristics, and thus simulate the dynamic life cycle inventory inputs. Further, multiple LCI databases and LCIA methods are available in this computational platform to create a flexible user-defined LCA system model, and the implications of varying combinations of inventory databases and impact assessment methods can also be quantified. Finally, by coupling with uncertainty analysis and sensitivity analysis, this quantitative LCA platform provides rapid and robust LCA results of candidate products, analyzes the tradeoffs among productivity, economic benefits, and environmental impacts, while identifying the critical process parameters that can advance systems-scale sustainability as a whole.

Topics:

Environmental Supply Curves Highlight Relationship between Environmental Indicators, Cost, and Biomass for Energy in National-Scale Assessment

Conference: Bioenergy Sustainability Conference

Year: 2019

Proceeding: 2019 Bioenergy Sustainability Conference

Group: Abstract Submissions

Session: Advances in integrated assessments of bioenergy sustainability

Authors:

Efroymson, R., Oak Ridge National Laboratory
Langholtz, M., Oak Ridge National Laboratory
Hilliard, M., Oak Ridge National Laboratory
Henriette, J., Oak Ridge National Laboratory

Bioenergy has the potential to be profitable while at the same time providing sustainability benefits. We present relationships between potential agricultural and forest biomass supply and environmental effects from a national-scale study of potential biomass for energy and bioproducts. We also develop new methods to visualize environmental effects, as well as tradeoffs or synergies between price and environmental effects, as biomass supply increases across the US. The visualizations are termed â€œenvironmental supply curves.â€ We conduct a county-level analysis of biomass scenarios and environmental effects of the potential biomass at the farmgate or roadside following harvest and prior to transport. Supply that could potentially be available annually in the US by 2040 at $66 per dry Mg is considered. Environmental supply curves for modeled indicators of water quality and quantity, GHG emissions, air quality, and biodiversity are presented. We visualize the environmental effects of a 2040 US biomass production scenario by environmental indicator, feedstock, region, and price. Biomass that meets or exceeds specified environmental or cost thresholds across the US can be quantified. For some indicators, the highest-cost biomass has rather benign environmental effects, whereas for other indicators, the highest-cost biomass has the most-adverse effects. The simulations of environmental effects of biomass highlight the importance of the specific context of the simulation, i.e., the particular indicator, feedstock, yield, region and supply chain step.

Navigating Tradeoffs Among Pretreatment Methods for Lignocellulosic Biomass of Varying Composition

BIOENERGY
SUSTAINABILITY
CONFERENCE

Conference: Bioenergy Sustainability Conference

Year: 2019

Proceeding: 2019 Bioenergy Sustainability Conference

Group: Abstract Submissions

Session: Advances in integrated assessments of bioenergy sustainability

Authors:

Li, Y., University of Illinois at Urbana-Champaign
Guest, J., University of Illinois at Urbana-Champaign

With sources including grass, wood, energy crops (e.g., *Miscanthus*), and various agrowastes (e.g., corn stover), lignocellulosic biomass is arguably the most abundant terrestrial biomass and has attracted significant attention as a promising feedstock for the emerging bioeconomy. It is primarily composed of lignin, hemicellulose, and cellulose, which are strongly interlinked thus highly resistant to (bio)chemical conversion, requiring effective pretreatment methods to break down the recalcitrant structure and liberate individual components prior to further processing.

Although a wide range of physical, chemical, physiochemical, and biological pretreatment methods have been investigated in the past decades, interactions between composition of lignocellulosic biomass and efficiency of pretreatment methods have not been examined in a quantitative and dynamic manner. In particular, several studies have recognized impacts of lignin content on pretreatment efficiency, yet no study has examined these impacts in a systematic way. This research quantitatively evaluated the impacts of lignin content on pretreatment efficiencies and categorizes prevailing pretreatment methods based on these impacts and mechanisms of the pretreatment methods. Connections between lignin content and pretreatment efficiency are incorporated into a biorefinery modeling platform where techno-economic analysis (TEA) is conducted to evaluate tradeoffs between efficiency and cost of pretreatment methods for lignocellulosic biomass of varying composition. Leveraging these results, the full spectrum of biomass composition is divided into individual zones based on the optimal pretreatment method and cost drivers for each pretreatment method are identified to prioritize future research.

Prioritization of Lignocellulosic Biomass Conversion Pathways for Economic and Environmental Sustainability

BIOENERGY
SUSTAINABILITY
CONFERENCE

Conference: Bioenergy Sustainability Conference

Year: 2019

Proceeding: 2019 Bioenergy Sustainability Conference

Group: Abstract Submissions

Session: Advances in integrated assessments of bioenergy sustainability

Authors:

Ki, D., University of Illinois at Urbana-Champaign
Guest, J., University of Illinois at Urbana-Champaign
Jin, Y. S., University of Illinois at Urbana-Champaign
Rao, C. V., University of Illinois at Urbana-Champaign

Production of chemicals (or bioproducts) from biomass provides a promising opportunity to enhance the economics and environment sustainably, compared to fossil-derived chemicals. It is expected and reported the increase of market size for bio-based chemical production. Bioproduct conversion technologies, however, are still in their early stages including strain development, optimization/scaling up of fermentation, and separation/purification of bioproducts. Of various possibilities in lignocellulosic biomass conversion processes, it is critical that we quickly prioritize conversion technologies and bioproduct targets. Thus, the research objective is to rapidly prioritize research for the valorization of lignocellulosic biomass via biocatalyzed conversions and establish a roadmap for process development. To achieve the objective, we firstly focus on sugar-derived building organic acids including lactic, citric, succinic, 3-hydroxypropionic, muconic, adipic, butyric, and propionic acids as bioproducts to characterize the economic performance with a rapid, rigorous simulation tool called Biorefinery Simulation and Techno-Economic Analysis Moduels in Python (BioSTEAM). Fermentation and separation unit processes are modeled based on experimental data and heuristic approaches for techno-economic analysis ($ kg^{-1}). The sustainability assessment via TEA and/or life cycle assessment (LCA) will leverage to prioritize research needs to achieve more rapid progress financially viable and environmentally sustainable products.

Recent LCA Studies Illustrate the Need for Improved Information on Impacts Associated with Changes in Belowground and Soil Carbon in Willow Biomass Crops

BIOENERGY SUSTAINABILITY CONFERENCE

Conference: Bioenergy Sustainability Conference

Year: 2019

Proceeding: 2019 Bioenergy Sustainability Conference

Group: Abstract Submissions

Session: Advances in integrated assessments of bioenergy sustainability

Authors:

Volk, T. A., SUNY College of Environmental Science and Forestry
Therasme, O., SUNY College of Environmental Science and Forestry
Fortier, M. O., University of California
Yang, S., SUNY ESF
Eisenbies, M., State University of New York College of Environmental Science and Forestry

Empirical data is lacking on belowground stored carbon and changes in soil carbon over time in willow biomass systems, especially when this involves a transition from different land uses into willow. Recent life cycle assessment (LCA) studies have shown the importance of these factors on the life cycle greenhouse gas emissions of willow biomass crops. A spatial LCA model in Central and Northern New York indicates that 92% of 9,718 suitable parcels for willow biomass production had negative GHG emissions. The average life cycle GHG emissions in the region were -126.8 kg CO_2 eq Mg^{-1} biomass on previous cropland or pasture land and ranged from -53.2 kg to -176.9 kg CO_2 eq Mg^{-1} across the region. However, for grassland converted to willow biomass crops, there was a modeled decrease in soil organic carbon (SOC) resulting in a slightly positive (27.7 kg CO_2 eq Mg^{-1} biomass) GHG balance. Uncertainty analysis showed large variations of probability distributions of GHG emissions in the five counties arising from differences in land use change, willow crop yield, and transportation distance. A LCA of willow biomass being converted to ethanol using hot water extraction and biological conversion also showed that belowground carbon storage and prior land use influenced on whether these systems had negative or positive emissions. Improved data on soil carbon dynamics and the belowground portion of willow biomass crops is needed to resolve this uncertainty.

Techno-Economic and Environmental Assessment of Phenolic Resins Production Using Lignin Byproducts from Biorefineries - a Case Study of 100% Phenol Replacement

BIOENERGY SUSTAINABILITY CONFERENCE

Conference: Bioenergy Sustainability Conference

Year: 2019

Proceeding: 2019 Bioenergy Sustainability Conference

Group: Abstract Submissions

Session: Advances in integrated assessments of bioenergy sustainability

Authors:

Wu, N., North Dakota State University (NDSU)
Pourhashem, G., North Dakota State University
Hussain, M., University of Haripur
Nejad, M., Michigan State University

Phenol formaldehyde (PF) resin is widely used in wood panel industry as an adhesive (more than 3 million metric tons per year globally), due to its unique properties (resistance to water and temperature, chemical durability, and bonding strength). Environmental impacts, health concern, and oil price fluctuations have motivated using renewable materials in PF resin production. Lignin, as the main byproduct from the paper industry and 2-G bioethanol biorefineries, is a promising candidate for petroleum-based phenol substitution in PF resin production for its exceptional phenyl propane units. Currently, less than 5% of the lignin is marketed for formulations of chemicals such as surfactants, adhesives, and anti-oxidants, and the remaining is utilized to generate energy through co-generation systems. Thus, the use of lignin by-product for value-added products such as lignin-formaldehyde (LF) resin can be substantial in promoting the economic and environmental benefits for biorefineries. In this study, we conducted a comparative techno-economic and environmental assessment on the developed LF resins, where lignin replaces 100% of phenol, and a commercially formulated phenol resorcinol formaldehyde (PRF). Industrial-scale process simulation was performed in Aspen Plus based on experimental data, while life cycle burdens were modeled using the TRACI 2.1 impact assessment method in SimaPro. The comparison showed an overall reduction in most environmental impacts for the lignin route, yet the process economics revealed higher cost for this bio-based resin compared to petroleum-based resin. Additionally, the alternative pathways of lignin utilization such as combustion, fast pyrolysis and depolymerization were discussed for economic impacts on the biorefineries.

The Potential for Eucalyptus Spp. Production in the U.S, Sustainability Considerations

Conference: Bioenergy Sustainability Conference

Year: 2019

Proceeding: 2019 Bioenergy Sustainability Conference

Group: Abstract Submissions

Session: Advances in integrated assessments of bioenergy sustainability

Authors:

Davis, M., Oak Ridge National Laboratory
Langholtz, M., Oak Ridge National Laboratory
Kainer, D., Oak Ridge National Laboratory

The technical viability of aviation biofuels has been demonstrated and a 29-billion-liter annual market exists, but feedstock price and availability remain a barrier. One potential biofuel feedstock, *Eucalyptus* spp., can produce an energy-dense terpene suitable for high-density synthetic hydrocarbon-type fuel (grade JP-10) production as well as provide substantial cellulosic-based feedstock for traditional jet fuels (e.g., Jet A aviation turbine fuel). We present on the results of an economic analysis of the potential for eucalyptus to fulfill US biojet and military-grade fuel markets in the near term (20 years) using the Policy Analysis System (POLYSYS). This study will also address the potential sustainability considerations associated with the modeled cold-tolerant variety of eucalyptus in an annual short-rotation woody crop coppice system across the Southeast and Gulf states. Our simulations focused on planting of genetically modified eucalyptus with potential to produce 14.5 dry Mg ha^{-1} yr^{-1}. Total eucalyptus production in primarily the U.S. Gulf States was modeled to potentially produce 51.4 million L yr^{-1} JP-10-type fuel, while also allowing 77.3 million L Gas and 75.4 million L JetA from total cellulosic material (woody & spent leaves) under a \$110 Mg^{-1} scenario within 10 years. Lignin could be used for heat and power in the production process, reducing steam-distillation and other energy costs, or as a soil amendment. The total estimated value of all products is \$493.3 million at a feedstock cost (farmgate) of \$99.7 million. Applicable sustainability indicators, land use considerations and other variables (e.g., water use) will be explored.

BiogasdonerightTM: Sustainable Energy to Enable Sustainable Agriculture

BIOENERGY
SUSTAINABILITY
CONFERENCE

Conference: Bioenergy Sustainability Conference

Year: 2019

Proceeding: 2019 Bioenergy Sustainability Conference

Group: Abstract Submissions

Session: Integrating systems Designing sustainable bioenergy systems

Authors:

Dale, B. E., Michigan State University

Both food and energy provisioning must expand significantly in the next few decades to meet the needs of a growing global population. However, if we expand either energy or food production using current practices, further environmental deterioration is assured from energy production and farming will continue to be both unprofitable for the farmers and environmentally destructive. We need rapidly scalable, proven technologies to expand sustainable food and energy production. Biogasdoneright™ is one such scalable, proven technology now being implemented at large scale in Italy. Biogasdonerightâ„¢ (or BDR) is an integrated set of agronomic practices that links biogas production with sustainable agriculture.

Under current bioenergy approaches, farmers do not benefit much, and, at least in part, the result has been very slow growth in bioenergy production. The Italian experience with BDR will be described along with the significant environmental and economic benefits that accrue to the farmers who practice BDR. During its rapid growth phase, Italian biogas production achieved exponential growth with a doubling time of 14 months. If agriculturally-based renewable energy is to advance rapidly, farmers must benefit significantly. If the proper incentives are put in place, BDR can expand rapidly in other countries beyond Italy, thereby greatly improving the sustainability of both energy and food provisioning. We urgently need to rethink how we incentivize and integrate sustainable agriculture and sustainable energy.

Development of a Bioenergy Sustainability Tradeoffs Assessment Resource (BioSTAR)

BIOENERGY SUSTAINABILITY CONFERENCE

Conference: Bioenergy Sustainability Conference

Year: 2019

Proceeding: 2019 Bioenergy Sustainability Conference

Group: Abstract Submissions

Session: Integrating systems Designing sustainable bioenergy systems

Authors:

Parish, E., Oak Ridge National Laboratory
Kline, K., Oak Ridge National Laboratory
Hilliard, M., Oak Ridge National Laboratory
Efroymson, R., Oak Ridge National Laboratory

Oak Ridge National Laboratory is developing a web-based visualization platform to inform decision-making for sustainable US cellulosic biomass production. The Bioenergy Sustainability Tradeoffs Assessment Resource (BioSTAR) will help the US Department of Energyâ€™s BioEnergy Technologies Office, industry and academic researchers to quantify environmental and socioeconomic costs and benefits and analyze potential tradeoffs associated with cellulosic feedstock choices in different regions of the country. By helping users to tailor their sustainability assessments to local stakeholder priorities, BioSTAR aims to identify bioenergy feedstocks and management practices that will maximize environmental and socioeconomic benefits to society (e.g., improvement of water and soil quality, a net increase in rural jobs) while minimizing negative impacts. The BioSTAR prototype has been designed with datasets and stakeholder feedback gathered during three case studies: (1) East Tennessee switchgrass-for-ethanol production, (2) Southeastern US bioenergy wood pellet production, and (3) Iowa corn stover and perennial grass production for ethanol using landscape design principles. These case studies have been used to develop documented procedures and tools to help users quantify and visualize effects of bioenergy options in a manner that provides a more complete picture of tradeoffs involved in complex human-biological systems. Over the next two years, BioSTARâ€™s capabilities will be expanded to allow users to evaluate the sustainability of their own bioenergy projects using either â€œdefaultâ€ national-scale datasets provided within the tool or their own datasets. Gathering feedback on BioSTARâ€™s utility and ease of use will help ORNL to refine and improve this adaptive management tool for future decision makers.

Empirical Evidence for the Climate Benefits of Decarbonizing US Automotive Transport with Liquid Vs Electric Bioenergy and Carbon Capture and Storage

BIOENERGY
SUSTAINABILITY
CONFERENCE

Conference: Bioenergy Sustainability Conference

Year: 2019

Proceeding: 2019 Bioenergy Sustainability Conference

Group: Abstract Submissions

Session: Integrating systems Designing sustainable bioenergy systems

Authors:

Hamilton, S. K., Michigan State University
Gelfand, I., Ben-Gurion University of the Negev
Kravchenko, A. N., Great Lakes Bioenergy Research Center
Jackson, R. D., Univ. of WIsconsin
Thelen, K. D., Great Lakes Bioenergy Research Center
Robertson, G. P., Great Lakes Bioenergy Research Center

Climate mitigation scenarios that limit global temperature increases to 1.5 Â°C rely on decarbonizing vehicle transport with bioenergy production combined with carbon capture and storage (CCS), but climate impacts for producing different bioenergy feedstocks have not been directly compared experimentally nor for alternative end uses. In a field experiment at two Midwest US sites on contrasting soils, yields of seven different feedstocks including maize stover, switchgrass, miscanthus, poplar, native grasses, early successional vegetation, and restored prairie varied substantially within sites but little between sites or among years. When used without CCS to power internal combustion vehicles, greenhouse gas emission intensities ranged from 17 to -183 g CO_2e MJ^{-1}in the order maize stover >> miscanthus â‰ˆ switchgrass â‰ˆ native grasses â‰ˆ poplar > early successional â‰¥ restored prairie; direct climate benefits ranged from 80% reductions in CO_2e compared to petroleum for stover to 130% for restored prairie, and were similar for electric vehicles without CCS. Used with CCS, reductions in emission intensities for ethanol powered vehicles ranged, on average, from 180 to 230% and for electric vehicles, from 330 to 410%. Energy production

potentials reflected yields and were in the order miscanthus > poplar > switchgrass > native grasses â‰ˆ maize stover > restored prairie â‰ˆ early successional. Extrapolation based on expected U.S. transportation energy needs suggests that negative emissions from bioenergy with CCS for light vehicle use could capture up to 2.2 Gt CO_2e yr^{-1}.

Environmental Impacts and Cost Analysis of Municipal Wastewater Treatment Anaerobic Digestion Expansion Options

BIOENERGY
SUSTAINABILITY
CONFERENCE

Conference: Bioenergy Sustainability Conference

Year: 2019

Proceeding: 2019 Bioenergy Sustainability Conference

Group: Abstract Submissions

Session: Integrating systems Designing sustainable bioenergy systems

Authors:

Ma, X., U.S. Environmental Protection Agency

Communities throughout the U.S. are shifting from the paradigm of waste disposal to viewing â €œwasteâ€ as a resource. Several industries are vying for the opportunity to utilize abundant organic waste streams. The presented research uses life cycle assessment and life cycle cost analysis to evaluate a Massachusetts wastewater treatment facility (WWTF), treating 23.5 million gallons per day, as it upgrades for resource recovery. The facility recently installed additional anaerobic digestion (AD) and combined heat and power (CHP) capacity, allowing acceptance of source separated organic (SSO) waste, boosting biogas production, and avoiding traditional food waste disposal. The planned capacity of the AD project will allow the facility to accept 40% of the current organic material available in the greater Boston region. We also analyze composting through windrow and aerated static pile (ASP) systems as an alternative disposal pathway for food waste. Co-digestion and composting are then compared to legacy food waste treatment options such as landfill and waste-to-energy combustion. The WWTF inventory data were developed using plant records and GPS-Xâ„¢ modeling software. Scenario and sensitivity analyses assess the effect of SSO acceptance rate, AD performance, avoided disposal processes, and cost parameters on results. Sensitivity results demonstrate the benefit of food waste co-digestion as compared to disposal via landfilling or waste-to-energy combustion. Food waste composting requires much lower capital costs compared to co-digestion, but is not able to capture the energy potential and associated environmental benefit of food waste energy recovery through co-digestion.

Impacts on Water Quality of Integrating Landscape Design with Conservation Practices for Biomass Production

BIOENERGY SUSTAINABILITY CONFERENCE

Conference: Bioenergy Sustainability Conference

Year: 2019

Proceeding: 2019 Bioenergy Sustainability Conference

Group: Abstract Submissions

Session: Integrating systems Designing sustainable bioenergy systems

Authors:

Ha, M., Argonne National Laboratory
Wu, M., Argonne National Laboratory

This study analyzes the effects on water quality of landscape design integrated with conservation practices for biomass production, as a part of a multi-institute project that identifies opportunities for growing biomass with profitability and environmental sustainability. The work assesses how and to what degree nitrogen, phosphorus, and suspended sediments runoff can be reduced at various scales across a watershed. A pair of Soil Water Analysis Tool models were developed, calibrated, and validated with 20-year hydrology and water quality records for the South Fork Iowa River (SFIR) and the Headwaters of North Raccoon River (HNRR) watersheds in Iowa. The models were used to simulate a landscape design scenario in which a portion of low- or negative-productivity land is converted to grow switchgrass for feedstock. We further simulated conservation practices and management such as riparian buffers and cover crop applications when stover is harvested at a rate of 30%, 45%, and 70%. Results demonstrated promising improvement of water quality under these scenarios. Compared with the historical baseline, when switchgrass is grown in low- or negative-profit land, 6%â€“12% of nitrogen, phosphorus, and sediments loss could be avoided in the SFIR and HNRR watersheds. Applying rye as a cover crop could reduce 15%â€“30% of the loadings, and installing riparian buffers could reduce up to 17% of nitrogen, 37% of phosphorus, and 71% of sediments. The two watersheds responded to the scenarios differently. Across all scenarios, the SFIR watershed showed responses more pronounced in nitrogen reduction than the HNRR watershed.

The Feasibility of Renewable Natural Gas in New Jersey

Conference: Bioenergy Sustainability Conference

Year: 2019

Proceeding: 2019 Bioenergy Sustainability Conference

Group: Abstract Submissions

Session: Integrating systems Designing sustainable bioenergy systems

Authors:

Bates, J., PSEG Institute for Sustainability Studies
Chandra, B., PSEG Institute for Sustainability Studies
Dyer, A., PSEG Institute for Sustainability Studies
Galindo, J., PSEG Institute for Sustainability Studies
Tran, C., PSEG Institute for Sustainability Studies
Olivier, V., PSEG Institute for Sustainability Studies

To determine the feasibility of renewable natural gas (RNG) in New Jersey, the following was achieved: development of financial and environmental models for different biogas resources in New Jersey, specifically landfills and wastewater treatment plants (WWTPs), determination of the effect of RNG on GHG emissions at different levels of penetration, analysis of the different incentives that are needed to drive the RNG market in NJ, research of successful RNG projects around the United States, and validation of the most beneficial way to implement RNG in New Jersey. The data show that the implementation of RNG has the potential to reduce the amount of GHG emissions and if properly structured with incentives could contribute meaningfully to the Clean Energy Economy of New Jersey. The financial and environmental models for landfills and WWTPs were compared to demonstrate the most beneficial implementation of RNG in New Jersey. Given similar incentives, the average profit margin of landfills greatly exceeds that of WWTPs by tenfold, in addition to the payback period for landfills being half of the payback period for WWTPs. The analysis indicates that landfills are found to be more beneficial for RNG production compared to WWTPs. Environmentally, landfills have larger RNG production compared to WWTPs, therefore, landfills will have larger GHG emissions reductions. The RNG potential of landfills greatly exceeds that of WWTPs when comparing the total amount of homes that can be heated using RNG: 89,500 homes for landfills and 21,500 homes for WWTPs.

Upgrading Biorefinery Lignin to Specialty Polymers: Environmental Performance and Economic Feasibility

Conference: Bioenergy Sustainability Conference

Year: 2019

Proceeding: 2019 Bioenergy Sustainability Conference

Group: Abstract Submissions

Session: Integrating systems Designing sustainable bioenergy systems

Authors:

Pourhashem, G., North Dakota State University
Wu, N., North Dakota State University

Ethanol production through bioconversion of lignocellulosic materials utilizes cellulose and hemicellulose of the biomass and leaves most of the lignin part as byproduct. Due to the complexity of lignin structure, and lack of commercialized pathways for converting it to higher-value products, many current lignocellulosic biorefinery designs assume using lignin-rich byproduct for energy to reduce the facilityâ€™s power needs. However, upgrading lignin to higher-value products not only can potentially boost the economic viability of a biorefinery, but it may also improve its environmental performance.

The phenolic structure and hydroxyl groups of lignin provide an advantage for lignin to be considered for replacing polymeric materials. We investigate the life cycle cost and environmental performance of using lignin in polymers with applications in coatings and composites. Lignin-based resins have displayed similar desirable characteristics to petroleum-based resins including comparable adhesion. Additionally, when used as a filler in composites, lignin can provide comparable mechanical properties. For commercialization decision, however, analyzing the environmental and cost advantage of these lignin-based materials is required.

We study the lignin-based resinsâ€™ supply chain and use process simulation to estimate the cost and environmental impacts of scaled-up lignin conversion to two types of resin and one filler for coating and composite applications. We compare these scenarios to lignin combustion for energy recovery. We demonstrate that in majority of the investigated environmental impact categories, the lignin-based materials show better performance than the commercial product in the market. However, for a more informed decision-making, the cost and environmental tradeoffs need to be considered.

From Bench to Booming: Deploying Engine-Efficiency Enhancing Emerging Bio-Blendstocks at a Large Scale

Conference: Bioenergy Sustainability Conference

Year: 2019

Proceeding: 2019 Bioenergy Sustainability Conference

Group: Abstract Submissions

Session: Keynote Talks

Authors:

Dunn, J., Argonne National Laboratory
Newes, E., National Renewable Energy Laboratory
Cai, H., Argonne National Laboratory
Zhang, Y., National Renewable Energy Laboratory
Brooker, A., National Renewable Energy Laboratory
Biddy, M., National Renewable Energy Laboratory

What if we designed fuels and engines cooperatively such that engines could exploit unique fuel properties towards higher efficiency? And what if these co-optimized fuels derived from biomass? These questions have been at the heart of the Department of Energy Bioenergy Technology Officeâ€™s Co-Optima initiative, which has explored the fuel properties of hundreds of compounds that could be made from biomass and the cost and environmental effects of selected bio-blendstocks. When a bio-blendstock is identified that has desirable fuel properties such as high octane, its pathway to wide-scale deployment must contend not only with going from bench to commercial scale production, but also challenges associated with fuel distribution and infrastructure. Additionally, the energy and environmental effects of bio-blendstocks deployed at a large scale must be evaluated. We have developed methods to combine vehicle choice, bioenergy sector, and life cycle analysis models to evaluate how vehicles with engines that are co-optimized with biofuel blends may emerge onto the market, driving demand for these blends which spurs the growth of the biorefinery industry and influences the energy and environmental effects of the light duty transportation sector. In addition, the analysis has considered the influence on net jobs as the biofuels industry grows. Two bio-blendstocks were considered in the analysis, a methyl furan mixture and isopropanol, which were compared against an ethanol as a baseline bio-blendstock. Analysis results will be presented along with a consideration of how to evaluate the effect of such a deployment of biofuel on land use change.

Realizing the Ecosystem Benefits of Cellulosic Biofuels: Insights from Integrated Economic-Ecosystem Models

BIOENERGY SUSTAINABILITY CONFERENCE

Conference: Bioenergy Sustainability Conference

Year: 2019

Proceeding: 2019 Bioenergy Sustainability Conference

Group: Abstract Submissions

Session: Keynote Talks

Authors:

Khanna, M., University of Illinois at Urbana-Champaign

Cellulosic biofuels, particularly from high yielding dedicated energy crops, are appealing as a renewable fuel because they have low carbon intensity, potential to be grown productively on marginal land and to provide other ecosystem benefits such as soil carbon sequestration and reduced nitrogen leaching. High-resolution satellite imagery and spatially resolved ecosystem models can quantify the potential for achieving these ecosystem benefits. Realizing these benefits will also depend on economic and policy incentives to produce cellulosic biofuels using energy crops and will vary spatially with crop yield and its riskiness, opportunity cost of land

and behavioral preferences of farmers. This talk will present insights from integrated economic-ecosystem modeling on the opportunities and challenges for realizing the promise of cellulosic biofuels from energy crops and areas for further research.

A Data-Driven Approach to Assessing Spatially-Explicit Air Emissions of Biomass Feedstock Production

BIOENERGY SUSTAINABILITY CONFERENCE

Conference: Bioenergy Sustainability Conference

Year: 2019

Proceeding: 2019 Bioenergy Sustainability Conference

Group: Abstract Submissions

Session: Research frontiers for sustainable bioenergy

Authors:

Inman, D., NREL
Heath, G., National Renewable Energy Laboratory (NREL
Zhang, Y., National Renewable Energy Laboratory
Hanes, R., National Renewable Energy Laboratory

Increased demand for biomass feedstock for biofuel production could lead to additional emission sources and may impact air quality. The National Renewable Energy Laboratoryâ€™s Feedstock Production Emissions to Air Model (FPEAM) is a tool, which can be used to develop county-level emission inventories of direct air pollutant emissions produced by the biomass feedstock supply chain in the United States. Here we provide an overview of FPEAM functionality and discuss several applications based on the newly released model.

FPEAM is a modular model, in which each module contains calculations for an activity type, except for the feedstock transportation and on-farm equipment operation modules which connect to the U.S. Environmental Protection Agencyâ€™s Motor Vehicle Emissions Simulator (MOVES) and NONROAD models, respectively. User-defined model inputs (e.g., equipment horsepower) or default data for certain scenarios are used to calculate inventories of volatile organic compounds, particulate matter, ammonia, nitrogen oxides, sulfur oxides generated by each activity in the biomass production and supply chain. The data-driven aspect of FPEAM means that the input data for a scenario (e.g.,production scale) rather than the model itself drives the type and magnitude of the pollutants. Modules within FPEAM can be turned on or off manually to adjust the scenario scope. Default data packaged with the FPEAM code can be used to calculate inventories of aforementioned pollutants generated by agricultural and forestry equipment operation, chemical application, and feedstock transportation. User-defined data can be used in place of the default data to analyze scenarios with alternate feedstocks and supply chain activities.

Autolysin Gene Deletion on Clostridium Saccharoperbutylacetonicum N1-4 Increase Strain Stability and Production for Biobutanol Fermentation

BIOENERGY SUSTAINABILITY CONFERENCE

Conference: Bioenergy Sustainability Conference

Year: 2019

Proceeding: 2019 Bioenergy Sustainability Conference

Group: Abstract Submissions

Session: Research frontiers for sustainable bioenergy

Authors:

Jimenez-Bonilla, P., Auburn University
Wang, Y., Auburn University
Gonzalez-de-Bashan, L., Auburn University
Blersch, D., Auburn University
Wang, Y., Auburn University

Biobutanol production is one of the most promissory alternatives to replace fossil fuels, because of its superior properties, such as: higher energy content, miscibility and hydrophobicity. Bio-butanol is an important chemical as well, used in the industry of paintings, farmaceuticals, cosmetics, food and others. Biotechnological production of butanol generate lower concentration, productivity and yield than ethanol production. Metabolic engineering on butanol-producer microorganisms is an opportunity to create a bigger economical stimulus for butanol industry. *C. saccharoperbutylacetonicum* N1-4 is a hyper-butanol producer strain, and generates a high titer (17 g/L of butanol). Besides the strain benefits, its applicability at industrial scale is limited by the strong autolytic behavior, which conducts to poor stability on continuous processes. The autolysis process makes cells self-destroy, and reduce the number of active microorganism, with a negative impact on the production. Also, the autolysins (enzymes responsible of autolysis) make the long-term continuous fermentation difficult and unstable. In this work, we have identified two autolysins and deleted them from the genome, in order to increase the strains stability, and production. Putative autolysins were identified from bioinformatic comparison of aminoacid homology in related strains. Then, putative genes had been overexpressed and biomass curves were compared with a control, to identify autolytic activity. Those genes fund autolytic were deleted using our customized CRISPR-Cas9 gene-editing tool, finding a better performance of the mutants, during fermentations.

Economic Potential for Using Marginal Land for Bioenergy Crop Production in the US

Conference: Bioenergy Sustainability Conference

Year: 2019

Proceeding: 2019 Bioenergy Sustainability Conference

Group: Abstract Submissions

Session: Research frontiers for sustainable bioenergy

Authors:

Debnath, D., University of Illinois
Chen, L., University of Illinois
Khanna, M., University of Illinois at Urbana-Champaign
Jiang, C., University of Illinios
Guan, K., University of Illinious at Urbana-Champaign

In the US, there are a significant increase in production of biofuels but mostly corn ethanol that have raised concerns regarding the diversion of food crops to biofuel production leading to increasing food prices and conversion of native grassland to cropland, commonly known as indirect land-use change effect. On the contrary, cellulosic biofuels derived from crop and forest residues, and energy crops produced on marginal land is less likely to compete with cropland. In this study, we examine the economic incentives for using different types of land (marginal and cropland) for energy crop production and the spatial variability in the pattern of land use. We aim to quantify the mix of energy crops and crop residues that will be produced at different biomass prices. We develop a new definition of â€œmarginal landâ€ based on historical land-use change pattern, which accounts socioeconomic, biophysical, and environmental factors. We then calculate available marginal land and its spatial distribution in the contiguous US and determine the optimal supply of feedstock under alternative biomass price structure by employing Biofuel and Environmental Policy Analysis Model. While previous studies that have assumed much larger availability of marginal land, our preliminary finding suggests that the amount of marginal land known with certainty is relatively small. Also, that marginal land is fairly productive, and so its opportunity costs are higher than assumed by previous studies, which may lead to the underestimation of the actual production costs of producing biomass.

Land Cover / Use Change Analysis: Data Sets, Nomenclature and Harmonization

Conference: Bioenergy Sustainability Conference

Year: 2019

Proceeding: 2019 Bioenergy Sustainability Conference

Group: Abstract Submissions

Session: Research frontiers for sustainable bioenergy

Authors:

Bard, S., Sharon K. Bard Consulting

A significant research area is the analysis of land cover/use change and its impact on carbon sequestration and ultimately on climate change and the sustainability of bioenergy. The challenges and criticism of this research often stem from the data sets used and the classification of the land units (such as acres or pixels) within the data sets. Many dimensions such as the nomenclature used for the land cover/use classes, the source of the data (e.g., remote sensing, surveys), the classification accuracies, and the research objectives impact the methodologies employed by the researchers. These dimensions have led to a diverse array of methods and a wide range of results, thereby generating a lot of debate about the validity of the data sets, research methods, and results. This paper attempts to move this research field forward by (1) summarizing the key issues related to the land cover/use data sets and their respective classifications used across the research area; (2) identifying implications of some of the methods used to address the issues; and (3) proposing approaches for the research community to consider for addressing the challenges encountered in using the currently available data sets.

Progress in the sustainability assessment of bio-ethanol production

BIOENERGY
SUSTAINABILITY
CONFERENCE

Conference: Bioenergy Sustainability Conference

Year: 2019

Proceeding: 2019 Bioenergy Sustainability Conference

Group: Abstract Submissions

Session: Research frontiers for sustainable bioenergy

Authors:

Manzini, F., National Autonomous University of Mexico

Considering our experience in the study of first-generation ethanol production from sugarcane in two modalities in Mexico, one from molasses and the other from direct juice, we compared the sustainability level of each modality in order to ï¬nd the modality with the highest sustainability. We developed a sustainability index integrated by six indicators applied in all the lifecycle stages of each crop modality, considering: greenhouse gas emissions (GHG), land use, water use, energy balance, ethanol production costs and the number of employments generated. Our research shows the production of sugarcane ethanol from direct juice has the highest sustainability index; therefore, it is the most sustainable modality. With the purpose to build a more powerful tool, and to expand it to other energy crops, we are assessing the need to increase the number of indicators up to twenty and to build a sustainability index applying a multi-criteria evaluation to estimate weight factors.

Rice Hulls Separation for Biofuels

BIOENERGY
SUSTAINABILITY
CONFERENCE

Conference: Bioenergy Sustainability Conference

Year: 2019

Proceeding: 2019 Bioenergy Sustainability Conference

Group: Abstract Submissions

Session: Research frontiers for sustainable bioenergy

Authors:

Lynam, J. G., Louisiana Tech University
Boldor, D., Louisiana State University
Kumar, N., Louisiana Tech University
Muley, P., Louisiana State University

Interest in â€œhomegrownâ€ sustainable biofuels has fluctuated directly with the price of non-renewable fuels. However, using potential food or feed, such as corn, to produce bioenergy seems unethical in a world of limited caloric resources. As a non-food renewable resource, lignocellulosic biomass from food processing could be used as an important starting point to produce biofuels indefinitely.

Rice hulls are a secondary agricultural residue that must be transported to processing centers with the rough grain for separation. Thus, costs for low density transportation for the biomass are negligible and unavoidable, unlike primary agricultural residues like corn stover. Rice hulls high silica content makes them unsuitable for animal feed, so they are often burned or landfilled. One way to valorize them is separation with deep eutectic solids (DES).

DES can separate rice hulls into a lignin-rich product with many potential uses, such as in solid fuel or in binders, as well as a cellulose-rich product with potential for enzymatic hydrolysis to produce sugars for fermentation to biofuels. We have separated these biofuels using DES.

Sustainable Algal Biofuel Production By Siting U.S. Algae Ponds Considering Seasonal Water-Stress Impact Using Aware-US

Conference: Bioenergy Sustainability Conference

Year: 2019

Proceeding: 2019 Bioenergy Sustainability Conference

Group: Abstract Submissions

Session: Research frontiers for sustainable bioenergy

Authors:

Lee, U., Argonne National Laboratory
Xu, H., Argonne National Laboratory
Hawkins, T. R., Eastern Research Group, Inc. (ERG)
Wang, M., Argonne National Laboratory
Coleman, A., Pacific Northwest National Laboratory
Wigmosta, M., Pacific Northwest National Laboratory
Sun, N., Pacific Northwest National Laboratory

Microalgae has great potential as biofuel feedstock, which can be produced in a diverse set of habitats. In this study, we employed the Algae Biomass Assessment Tool, an algae growth model, and the AWARE-US regional water-stress impact assessment method to screen potential sites, simulate algae growth and biomass productivity, and assess regional and seasonal water-stress impacts of open-pond freshwater algae production. AWARE-US characterization factors estimated based on measured U.S. freshwater supply and demand are used. In order to support sustainable decisions on siting of algae ponds, we have compared three algae ponds siting strategies based on (1) biomass productivity, (2) water-use efficiency (WUE), and (3) seasonal water-stress impact. Compared to the strategy only considering biomass productivity, water stress impact and total water consumption can be reduced by 97% and 50%, respectively, by utilizing AWARE-US, while only having <1.7% of reductions in algae productivity. This study shows that careful selection of algae ponds with consideration of freshwater supply and demand can enhance the sustainability of algal biofuel production. In particular, it shows considering seasonal water stress for algae-system deployment can help avoid sites with short but high seasonal water stress. Further investigation is needed to reduce the water stress even further by utilizing non-freshwater algal ponds.

2 Stage Anaerobic Digestion of Post Extracted Algae Residue (PEAR) of Algae to Generate Biohythane

BIOENERGY SUSTAINABILITY CONFERENCE

Conference: Bioenergy Sustainability Conference

Year: 2019

Proceeding: 2019 Bioenergy Sustainability Conference

Group: Poster Submissions

Session: Poster Session

Authors:

Abimbola, T., Please Select

The elemental composition of woody biomass has important consequences for both energy conversion processes and agronomic management of nutrient budgets. We evaluated the effects of genotype, environment and their interaction on elemental concentrations in shrub willow. Samples from six genotypes in four trial locations were separated into bark and wood, and analyzed for C, N and H by combustion and for elemental composition by ICP-OES. An additional set of whole biomass samples for the six genotypes from six additional trial locations were also analyzed, totaling 252 total independent observations. Generally *Salix purpurea* genotypes had high N, Mg and Fe concentrations in bark, whereas *S. miyabeana* genotypes had high Ca concentrations in both bark and wood. Hybrids between these two species displayed intermediate values in elemental composition, suggesting an additive mode of inheritance for these traits. Multivariate analyses of the broader suite of biomass samples showed strong environmental effects on elemental composition, but there were a number of significant genotype × environment interactions. Nutrient export rates at harvest were largely a function of yield, but some genotypes showed signs of luxury uptake when N was abundant. These results showed important genotype-specific preferences in nutrient uptake, which may need to be addressed in nutrient management plans. Our test sites did not include heavy metal contaminated soils, but these data suggest a capacity to exploit existing genetic resources for improving phytoremediation applications and for development of effective riparian buffer strips to reduce excess nutrient export.

A Stochastic Techno-Economic Analysis of a Forest Biomass Feedstock Supply Chain in the Northeast United States

Conference: Bioenergy Sustainability Conference

Year: 2019

Proceeding: 2019 Bioenergy Sustainability Conference

Group: Poster Submissions

Session: Poster Session

Authors:

Ha, H., State University of New York - College of Environmental Science & Forestry
Brown, T., State University of New York - College of Environmental Science & Forestry
Quinn, R. J., State University of New York - College of Environmental Science & Forestry
Volk, T. A., SUNY College of Environmental Science and Forestry
Malmsheimer, R., State University of New York - College of Environmental Science & Forestry
Fortier, M. O., University of California
Bick, S., bNortheast Forests, LLC
Frank, J., State University of New York - College of Environmental Science & Forestry

Utilizing low-carbon forest biomass feedstocks in the Northeast United States for bioenergy applications has a substantial potential to mitigate greenhouse gas emissions by replacing fossil fuels. Understanding the economic feasibility of the regional forest biomass feedstock supply chain with uncertain variables is critical to supply feedstock for bioenergy projects whose lifetimes typically span over decades. This study developed a stochastic techno-economic analysis (TEA) model to assess the economic feasibility of the regional supply chain by using three 24-year scenarios that would produce primarily clean chips and dirty chips mixed with forest co-products. The stochastic TEA model produced net present values (NPVs) and minimum selling prices (MSPs) by using Monte Carlo simulation. The 1st and 99th percentiles of 10,000 MSP outcomes for clean chips were $40.90 and $63.80 per green metric tonne (Mg) while those for dirty chips from two different scenarios were $29.20 and $45.90/Mg and $25.50 and $47.90/Mg. The probabilities of gain for all three scenarios fell between 55% and 65%. Our sensitivity analysis revealed the variations of the NPV outcomes were most sensitive to the uncertainty of predominant forest biomass feedstock production levels and delivered prices, followed by fuel consumption and price variables.

Anaerobic Digestion and Phosphorous Recovery from Wastewater Produced in a Lignocellulosic Ethanol Plant – a Techno-Economic Analysis Using Aspen PLUS

BIOENERGY SUSTAINABILITY CONFERENCE

Conference: Bioenergy Sustainability Conference

Year: 2019

Proceeding: 2019 Bioenergy Sustainability Conference

Group: Poster Submissions

Session: Poster Session

Authors:

Yang, S., University of Florida
Wu, N., University of Florida
Pullammanappallil, P., University of Florida
Svoronos, S., University of Florida

Lignocellulosic ethanol plants employing a biochemical platform produce large quantities of wastewater containing undegraded components of plant biomass, catalysts and coproducts from pretreatment, residual nutrients and microbial biomass from fermentation, and condensates from various unit operations. A techno-economic analysis, of a commercial plant employing dilute phosphoric acid pretreatment integrated with anaerobic digestion and phosphorus recovery processes for wastewater treatment, was conducted using a model developed on ASPEN Plus flowsheeting software. The model was calibrated using data from pilot plant and laboratory scale experiments. Biogas produced from anaerobic digestion could displace 40% of natural gas required for process heat. Due to lower capital costs, the breakeven cost of ethanol production (57.85 cents/liter) was comparable to that of a process that recovers undegraded fibers for combustion. The costs are further lowered (55.35 cents/liter) by recovery and sale of phosphorus. If incorporating anaerobic digestion of pretreated biomass to provide all the fuel requirements for plant, the cost increases to 61.67 cents/liter. This is an attractive scenario if a penalty for fossil fuel usage (carbon credits) is levied at current European levels.

Bioenergy Via Char Gasification with Carbon Capture and Storage: An Alternative to Beccs

BIOENERGY SUSTAINABILITY CONFERENCE

Conference: Bioenergy Sustainability Conference

Year: 2019

Proceeding: 2019 Bioenergy Sustainability Conference

Group: Poster Submissions

Session: Poster Session

Authors:

Chun, D., The Cooper Union
Zhao, P., Cooper Union
Davis, B. J., The Cooper Union
Simson, A., The Cooper Union for the Advancement of Science and Art
Antony, A., Pennslyvania State University

IPCC techno-economic models suggest that staying below the 2°C limit set forth by the Paris Agreement will require negative emissions technologies, with a majority of models incorporating bioenergy with carbon capture and storage (BECCS). However, there are several limitations to the most commonly described BECCS approach such as competition for water, land, and energy resources as well as difficulty in separating CO_2. We propose a novel BECCS alternative which uses biowaste and pyrolyzes it into biochar. This biochar would then be gasified at a larger power plant in CO_2 to generate syngas as a fuel for a solid oxide fuel cell (SOFC). The fuel cell would produce a stream which is almost all H_2O and CO_2, easily separated through condensation into a CO_2 stream for sequestration.

The biochars we have derived experimentally were from food waste (walnut shells, pistachio shells, corn cob, and pumpkin seeds). Several pyrolysis conditions such as ramp rate, moisture content, and purge atmosphere were investigated to understand reactivity in gasification. Chars with high ash content (e.g. corn cob) were most reactive with CO_2, perhaps from the inorganics present in ash. Fast pyrolysis yielded a more reactive biochar and biowaste drying before pyrolysis made the biochar less reactive in gasification. We will present both a techno-economic analysis of the alternative BECCS scheme and experimental data on pyrolysis and biochar gasification. We performed ASPEN simulations to compare our data to traditional BECCS models. [Fajardy, M; Dowell, N.Energy Environ. Sci., 2017, 10, 1389-1426].

Cellulose Dissolution and Biomass Pretreatment Using Quaternary Ammonium Ionic Liquids Prepared from G-, S-, and H-Type Lignin-Derived Benzaldehydes and Dimethylcarbonate

BIOENERGY
SUSTAINABILITY
CONFERENCE

Conference: Bioenergy Sustainability Conference

Year: 2019

Proceeding: 2019 Bioenergy Sustainability Conference

Group: Poster Submissions

Session: Poster Session

Authors:

Socha, A., Queens University of Charlotte

The synthesis of novel benzyl alkylammonium ionic liquids (IL) from starting materials derived from grass, softwood, and hardwood lignin was performed using batch reactor systems followed by facile methyl carbonate ion exchange chemistry. Approximately 90% yields of the ILs were achieved over three synthetic steps, and all novel materials were fully characterized by 1 H and 13 C nuclear magnetic resonance spectroscopy and high-resolution mass spectrometry. The structureâˆ’ activity relationships of the lignin-derived ILs were evaluated for their ability to dissolve microcrystalline cellulose and convert switchgrass to fermentable sugars (glucose and xylose). All ILs tested, including a 1:1:1 mixture of ILs prepared from vanillin, syringaldehyde, and p anisaldehyde yielded 52âˆ’ 71% of the glucose and 50âˆ’ 71% of the xylose from the total amount of glucan and xylan available in the raw feedstock. These data compared well to the yields obtained from switchgrass pretreatment with the IL, 1-ethyl-3- methyl imidazolium acetate, which gave 71% of the glucose and 63% of the xylose from the total glucan and xylan in the raw feedstock, respectively.

Comparison of Ultrasonication and Conventional Heating Methods for Biodiesel Production Using Calcium Oxide Catalysts Obtained from Calcined Shells of Eggs, Oyster, and Mussels

Conference: Bioenergy Sustainability Conference

Year: 2019

Proceeding: 2019 Bioenergy Sustainability Conference

Group: Poster Submissions

Session: Poster Session

Authors:

Chong, N. S., Middle Tennessee State University
Ooi, B. G., Middle Tennessee State Univerisity
Abdulramoni, S., Middle Tennessee State Univerisity
Perna, S., Middle Tennessee State University
McGruder, T., Middle Tennessee State University

There is an increasing demand for alternative fuels that are environmentally friendly because petroleum-based fuels contribute to the net production of carbon dioxide that has been attributed to the global warming trend. Besides, petroleum is considered a non-renewable resource that cannot be easily replenished like the agricultural crops for producing the biomass-derived fuels like biodiesel. In this study, biodiesel will be produced using both ultrasonic-assisted transesterification reaction using shell-derived calcium oxide catalyst as well as conventional heating of the reaction between soya bean oil and methanol in the presence of sodium hydroxide catalyst. The ultrasonication approach has been shown to produce high yield biodiesel esters in reaction time as short as 5-10 minutes. The efficiency of both methods will be compared in terms of their production cost, reaction yields, product selectivity, catalyst durability, and environmental factors. The emission profile of biodiesel blended with petroleum-based diesel and different oxygenated additiveswill be analyzed by Fourier Transform Infrared Spectrometry (FTIR) and gas chromatography coupled to mass spectrometry (GC-MS). The extent of the reduction in the emissions of carbon monoxide, volatile organic compounds, and particulate matter will be compared. The cold flow properties and fuel stability for the different diesel fuels blended with biodiesel and other biomass-derived fuel additives will also be evaluated. The synergistic performance of specific combinations of fuel additives and biodiesel blendstock will be emphasized.

Conversion of Sugarcane Bagasse to Bioenergy over a Metal Oxide Catalyst

Conference: Bioenergy Sustainability Conference

Year: 2019

Proceeding: 2019 Bioenergy Sustainability Conference

Group: Poster Submissions

Session: Poster Session

Authors:

Hao, N., University of Tennessee

Valorization of sugarcane bagasse to high-quality biofuel precursors using metal oxide catalysts was studied in this work. Preliminary results showed that a hydrocarbon selectivity of 14.9 wt% can be achieved in the upgraded sugarcane bagasse derived pyrolysis oils. Non-condensed phenolic hydroxyl groups were almost eliminated in the upgraded pyrolysis oils. A detailed performance study of the metal oxide catalysts was carried out for the reaction condition optimization (temperature, time, catalyst loading). Different solvents (e.g. ethanol, methanol, isopropanol, acetone) were tested in this upgrading reaction. A series model compound study was conducted to study the possible repolymerization and condensation in this catalytic upgrading reaction. ^{1}H, ^{31}P, HSQC NMM, GPC, and GC-MS were used for pyrolysis oil structural analysis for understanding the reaction pathways. A stability test of the catalyst and a detailed characterization of the fresh/used catalysts were performed and was carried out.

Deciphering Yeast Transcriptomics: A Unique Tool in Understanding Biochemical Basis for Ethanol Production

Conference: Bioenergy Sustainability Conference

Year: 2019

Proceeding: 2019 Bioenergy Sustainability Conference

Group: Poster Submissions

Session: Poster Session

Authors:

Nanjundaswamy, A., Alcorn State University
Mandyam, K., Alcorn State University
Kaptral, V., IgenBio
Vaisvil, B., IgenBio
Schmitt, D., IgenBio

Saccharomyces cerevisiae is so crucial to bioethanol production that its performance in ethanol production decides whether a process is economical or not. Biochemical basis of ethanol production appears simple with dosing of yeast to sugar rich medium where yeast produces ethanol through fermentation process. But a closer look gives a much complex process involved in ethanol production. Yeast will be growing in an unfavorable high solid medium in presence of inhibitors such as ethanol and organic acids. Also, there will be mass transfer limitations such as oxygen transfer rates (OTR) and temperature. In order to optimize the ideal conditions for the yeast growth and ethanol production it is critical to understand the yeast behavior at transcriptional level. The present study employed corn stover and other leading cellulosic biomasses for ethanol production in benchtop bioreactor. Samples were harvested at 24, 48 and 72h and analyzed for yeast behavior during fermentation. Transcriptional analysis of yeast indicated stress related genes up-regulation during fermentation. The presentation will discuss transcriptional information of important biochemical pathways as influenced by the fermentation conditions.

Designing a Sustainable Bioenergy System That Integrates Localized Fast Pyrolysis and Electrocatalytic Hydrogenation with Central Hydroprocessing for the Production of Gasoline-like Fuels: Techno-Economic and Life Cycle Analysis

Conference: Bioenergy Sustainability Conference

Year: 2019

Proceeding: 2019 Bioenergy Sustainability Conference

Group: Poster Submissions

Session: Poster Session

Authors:

Das, S., MSU
Saffron, C. M., Michigan State University

Valorization of sugarcane bagasse to high-quality biofuel precursors using metal oxide catalysts was studied in this work. Preliminary results showed that a hydrocarbon selectivity of 14.9 wt% can be achieved in the upgraded sugarcane bagasse derived pyrolysis oils. Non-condensed phenolic hydroxyl groups were almost eliminated in the upgraded pyrolysis oils. A detailed performance study of the metal oxide catalysts was carried out for the reaction condition optimization (temperature, time, catalyst loading). Different solvents (e.g. ethanol, methanol, isopropanol, acetone) were tested in this upgrading reaction. A series model compound study was conducted to study the possible repolymerization and condensation in this catalytic upgrading reaction. ^{1}H, ^{31}P, HSQC NMM, GPC, and GC-MS were used for pyrolysis oil structural analysis for understanding the reaction pathways. A stability test of the catalyst and a detailed characterization of the fresh/used catalysts were performed and was carried out.

Designing a Sustainable Bioenergy System That Integrates Localized Fast Pyrolysis and Electrocatalytic Hydrogenation with Centralized Hydroprocessing for the Production of Gasoline-like Fuels: Technoeconomic and Life Cycle Analyses

Conference: Bioenergy Sustainability Conference

Year: 2019

Proceeding: 2019 Bioenergy Sustainability Conference

Group: Poster Submissions

Session: Poster Session

Authors:

Das, S., MSU
Saffron, C. M., Michigan State University

According to the 2016 Billion Ton Report, the annual harvestable biomass in the U.S. (that can be produced at $60/ton or less) by the year 2030 is not sufficient to meet the annual energy demands of the U.S. transportation sector alone. Therefore, processes must be developed to upgrade the energy content of biofuels to match that of petroleum-based fuels. Such upgrading must also overcome the barriers resulting from biomassâ€™ dispersed nature and high costs of transporting a low bulk density feedstock. Depot-based fast pyrolysis partially overcomes these barriers by densifying biomass into bio-oilâ€"however, this mixture is reactively unstable and corrosive. Subsequent electrocatalytic hydrogenation (ECH) potentially stabilizes this bio-oil to a form that can be transported and stored at a centralized refinery for further upgrading to final fuel. ECH involves using mild temperature, atmospheric pressures, and renewable electricity, such as wind or solar, to split water on a catalytic cathode to produce *in situ* hydrogen and consequently perform the reduction of reactive aromatic compounds present in bio-oil. Localized fast pyrolysis and ECH, followed by centralized hydroprocessing has great potential as a sustainable bioenergy system for producing biofuels from biomass. Investigation of the systemâ€™s thermodynamics has revealed its advantages over traditional cellulosic ethanol fermentations in terms of energy and carbon yields. Therefore, full scale techno-economic and life cycle analyses were performed to quantify the environmental impacts and economic viability of the bioenergy system and ultimately determine its large-scale commercial application.

Genetic and Environmental Influences on First-Rotation Shrub Willow (Salix spp.) Bark and Wood Elemental Composition

Conference: Bioenergy Sustainability Conference

Year: 2019

Proceeding: 2019 Bioenergy Sustainability Conference

Group: Poster Submissions

Session: Poster Session

Authors:

Fabio, E., Cornell University
Smart, L., Cornell University

The elemental composition of woody biomass has important consequences for both energy conversion processes and agronomic management of nutrient budgets. We evaluated the effects of genotype, environment and their interaction on elemental concentrations in shrub willow. Samples from six genotypes in four trial locations were separated into bark and wood, and analyzed for C, N and H by combustion and for elemental composition by ICP-OES. An additional set of whole biomass samples for the six genotypes from six additional trial locations were also analyzed, totaling 252 total independent observations. Generally *Salix purpurea* genotypes had high N, Mg and Fe concentrations in bark, whereas *S. miyabeana* genotypes had high Ca concentrations in both bark and wood. Hybrids between these two species displayed intermediate values in elemental composition, suggesting an additive mode of inheritance for these traits. Multivariate analyses of the broader suite of biomass samples showed strong environmental effects on elemental composition, but there were a number of significant genotype × environment interactions. Nutrient export rates at harvest were largely a function of yield, but some genotypes showed signs of luxury uptake when N was abundant. These results showed important genotype-specific preferences in nutrient uptake, which may need to be addressed in nutrient management plans. Our test sites did not include heavy metal contaminated soils, but these data suggest a capacity to exploit existing genetic resources for improving phytoremediation applications and for development of effective riparian buffer strips to reduce excess nutrient export.

High Solids Roller Bottle Enzymatic Hydrolysis and Respirometer Fermentation for High Throughput Screening of Lignocellulosic Biomass

Conference: Bioenergy Sustainability Conference

Year: 2019

Proceeding: 2019 Bioenergy Sustainability Conference

Group: Poster Submissions

Session: Poster Session

Authors:

Chandrasekar, M., Michigan Technological University
Ong, R. G., Great Lakes Bioenergy Research Center
Sato, T., Great Lakes Bioenergy Research Center
Joshi, L., Great Lakes Bioenergy Research Center
Kreig, K., Michigan Technological University
Cronk, B., Michigan Technological University
Burke, E., Great Lakes Bioenergy Research Center
Debrauske, D. J., University of Wisconsin

High solids enzymatic hydrolysis increases the final fermentable sugar concentration in the hydrolysate, thereby making the process more economically viable by reducing equipment size and energy requirement. But this method has bottlenecks like improper mixing and low enzyme accessibility, which affects the final sugar concentration, which can be rectified through horizontal mixing on a laboratory scale roller bottle mixer. Ultimately the objective is to make this approach applicable to rapidly identify the effects of abiotic and biotic stresses on different feedstocks under industrially relevant process conditions. The samples were subjected to AFEX (ammonia fiber expansion) pretreatment at 1:1 ammonia loading for corn stover and 2:1 for switchgrass, sorghum, mixed prairie and miscanthus. Enzymatic hydrolysis was carried out on the pretreated samples in 85 mL polycarbonate centrifuge tubes at 6% and 9% glucan loading (approximately 25% and 30% by weight). The biomass was autoclaved prior to enzymatic hydrolysis to limit microbial contamination. An enzyme cocktail of 70% cellulase and 30% hemicellulase was used for these experiments, and the pH was controlled using phosphate buffer. After 72 hours of enzymatic hydrolysis at 50°C the samples were subjected to centrifugation at 12000 rpm for either 1 or 3 hours. The hydrolysates obtained from these experiments were separately subjected to fermentation by Saccharomyces cerevisiae andZymomonas mobilis. The hydrolysates and fermentation products were analyzed using HPLC to obtain sugar and ethanol concentrations respectively. Based on the results, 6% glucan loading was found to be more effective considering the final sugar and ethanol yields.

Hydrogen Production from Malate or Glycerol By Purple Non-Sulfur Photosynthetic Bacteria

Conference: Bioenergy Sustainability Conference

Year: 2019

Proceeding: 2019 Bioenergy Sustainability Conference

Group: Poster Submissions

Session: Poster Session

Authors:

Averitt, J., APSU
Markov, S., APSU

Replacement of fossil fuels with alternative fuels is imperative to counteract climate change and fossil fuel depletion. Molecular hydrogen (H2) is an attractive energy carrier as it does not contain carbon and does not produce harmful byproducts when utilized as a fuel. Hydrogen production of two species of photosynthetic non-sulfur purple bacteria,Rubrivivax gelatinosus and Rhodospirillum rubrum, were studied with malic acid or glycerol as sole carbon source under illumination in anoxygenic conditions. The results in test tubes showed that the maximal hydrogen photoproduction rate (196 mL/g DW/hour) by R. rubrum was obtained under 10% concentration of malic acid in the medium. For R.gelatinosus maximum hydrogen photoproduction rate (108 mL/g DW/hour) was observed in the presence of 10% malic acid in small scale test tubes. For the pilot scale of our study we selected R. rubrum with 10% malic acid to grow in a 17 L photobioreactor that was under artificial illumination with agitation by paddle stirring at 50 RPM. Bacterial growth was shown as an increase in the optical density. During the period of agitation of bacterial culture, there was no visible production of H_2. The agitation prevented hydrogen produced from being release from solution to the gas phase of the photobioreactor. The data suggested that it is possible to produce hydrogen by photosynthetic bacteria with inexpensive carbon sources such as malic acid in a pilot scale photobioreactor with additional energy savings (no agitation).

Incorporation of Agent-Based Modeling and Life Cycle Sustainability Assessment to Improve Assessment of Woody-Biomass Based Biofuels

Conference: Bioenergy Sustainability Conference

Year: 2019

Proceeding: 2019 Bioenergy Sustainability Conference

Group: Poster Submissions

Session: Poster Session

Authors:

Zupko, R. II, Pennsylvania State University
Rouleau, M., Michigan Technological University

Woody-biomass based biofuels are an important component to a viable renewable energy portfolio. However, it remains unclear if it is possible to produce sufficient feedstocks without introducing negative environmental and socio-economic consequences. Projections of feedstock supply and land-use change impacts are critical to bioenergy development given the large capital expenditures and time horizons associated with biofuel facilities. Our study demonstrates how agent-based modeling (ABM), a form of computer simulation, can be integrated with life cycle sustainability assessment (LCSA) to make such projections in a way that captures the complex feedback between individual land-use changes and broader socio-economic and environmental concerns. Our ABM is used to project the harvest behavior of thousands of individual forest owners and their resulting woody biomass feedstocks made available to loggers with resulting land-use change impacts then used to inform our life cycle inventory data for our LCSA. To illustrate the merits of our approach, we conducted a case study analysis of a proposed biorefinery facility in Ontonagon, MI. Our results show that, although Ontonagon is a promising biorefinery site, the social availability of woody biomass feedstocks may inhibit energy production while increased harvesting may also negatively impact regional wetlands. Ultimately, our case study demonstrates the applicability of our ABM-LCSA method and its added value to bioenergy planning and the development of sustainable biofuel facilities.

Life Cycle Analysis of a Biobased Resin: A Case Study on Acetoacetylated Lignin Resin

Conference: Bioenergy Sustainability Conference

Year: 2019

Proceeding: 2019 Bioenergy Sustainability Conference

Group: Poster Submissions

Session: Poster Session

Authors:

Pandey, R., North Dakota State University
Webster, D., North Dakota State University
Pourhashem, G., North Dakota State University

Lignin is the second most abundant biopolymer found on earth. Lignin is produced as a by-product by wood and paper industries, and to a lesser extent by cellulosic biorefineries. The bulk of the lignin generated is burnt as a low value fuel. In order to commercialize cellulosic biorefineries and promote biobased economy, cellulosic fuel needs to be cost competitive and have lower environmental impacts compared to their conventional counterparts. One way of sustaining biorefinery economically and commercializing it is to valorize the by-product lignin into higher value specialty products. However, before an upgrading pathway for lignin is recommended, a full environmental assessment is needed.

One attractive route for upgrading lignin is functionalizing it for coating applications with different unique properties. The aromatic and aliphatic hydroxyl functional groups in lignin can be substituted with acetoacetate functionality to produce acetoacetylated lignin resin. This resin can then be crosslinked with amines, polyisocyanates, or melamine-formaldehyde resins to obtain thermosets. The advantage of this process is that the polyol used in synthesis acts as a reactive diluent while crosslinking, which makes the process waste free.

In our study, we investigate the environmental performance of this lignin-based resin. We use the Life Cycle Assessment (LCA) method to assess the environmental impacts of the lignin-based resin and show whether the production of this resin is more sustainable than similar conventional coatings. The insights provided by the results from this study can provide valuable inputs in broadening the concept of integrated biorefineries.

Life Cycle Greenhouse Gas Emissions of Biofuels Production from Willow Hot Water Extract

Conference: Bioenergy Sustainability Conference

Year: 2019

Proceeding: 2019 Bioenergy Sustainability Conference

Group: Poster Submissions

Session: Poster Session

Authors:

Therasme, O., SUNY College of Environmental Science and Forestry
Volk, T. A., SUNY College of Environmental Science and Forestry
Fortier, M. O., University of California
Amidon, T., State University of New York

Hot water extraction (HWE), a pretreatment step incorporated in a multi-pathway biorefinery, yields valuable chemicals and sugars that can be used as feedstocks for bioethanol production and other products. A cradle-to-grave life cycle assessment is performed to assess the greenhouse gas (GHG) emissions associated with the production of 1 MJ of ethanol by fermentation of sugars derived from HWE of willow biomass. This study analyzes eight scenarios from the combination of two previous land uses (cropland and grassland), two harvest and storage seasons (summer and winter), and two biorefinery energy sources (biomass and natural gas). Ethanol produced from willow grown on cropland can sequester 0.012 kg $CO_{2\,eq}$ MJ^{-1} for a supply system incorporating winter harvest and using biomass to meet the energy requirement of the biorefinery. However, for the natural gas-based scenarios, the estimated GHG emissions for willow grown on cropland are 0.078 kg $CO_{2\,eq}$ MJ^{-1} and 0.118 kg $CO_{2\,eq}$ MJ^{-1} on grassland. The sensitivity analysis indicates that the life cycle GHG emissions associated with each scenario respond differently to changes in input parameter values; ethanol yield and HWE mass removal have greater influence on the net GHG emissions for natural gas than biomass based scenarios, storage duration is more influential for summer than winter harvests, and the fraction of suitable parcels converted into willow production has a higher impact for grassland than cropland. This study demonstrates that bioethanol produced from willow grown on cropland can be used as a mean of carbon sequestration while producing liquid transportation fuel.

Microalgal Proteins to Bioplastics: Techno-Economic and Life Cycle Assessment

Conference: Bioenergy Sustainability Conference

Year: 2019

Proceeding: 2019 Bioenergy Sustainability Conference

Group: Poster Submissions

Session: Poster Session

Authors:

Quiroz, D., Colorado State University

The low price of petroleum-based fuels and the high capital costs of algal growth architectures have made it challenging for microalgae-derived fuels to compete in the market. Over the last decade, researchers have been focusing on a variety of alternative co-products such as nutraceuticals, cosmetics, pharmaceuticals, and animal feed. Nonetheless, the quality of the protein plays an important role in the determination of the most suitable co-product. Given their large market size and relatively high value, bioplastics are a promising alternative for low quality protein biomass. This study explores the economic and sustainability implications of a bioplastic biorefinery with biofuels co-products. The model of our system includes the different sub-processes involved in the production of bioplastic feedstock and has been validated by experimental data. A total of nine scenarios were modeled, each with different levels of processing. A techno-economic and life cycle assessment were performed to determine the economic viability and environmental impact of bioplastic production. Results indicate significant improvements in environmental performance of the produced bioplastic feedstock, with reductions ranging between 67-116% compared to petroleum based plastic resins. Additionally, the minimum selling price of the bioplastic feedstock is within the range of economic competition with prices as low as $970 USD tonne^{-1}. These results indicate that an algae biorefinery focused on bioplastic feedstock production and fuels have significant potential to operate both economically and sustainably.

New Generation of Nanocatalysts for CO2 Electrochemical Conversion into Fuels

BIOENERGY
SUSTAINABILITY
CONFERENCE

Conference: Bioenergy Sustainability Conference

Year: 2019

Proceeding: 2019 Bioenergy Sustainability Conference

Group: Poster Submissions

Session: Poster Session

Authors:

Kumar, B., Elizabeth City State University

Electrochemical carbon dioxide reduction to fuels/valuable chemicals is a carbon-neutral process, contributing to sustainability and helping to mitigate potential climate change. If the process run by utilizing renewable energy, it can also be used to resolve intermittency issue associated with renewable energy (such as solar and wind energy) via storing energy into chemical bonds. Hence, it is a highly desirable goal to convert CO_2 into fuel precursors such as carbon monoxide, ethylene, methanol, or formic acid using renewable sources of energy (i.e. solar, geothermal, wind, etc.) as the energy input for the process, thereby presenting a convenient way to recycle CO_2 into fuels. The central problem with this approach, however, is the higher over-potential needed for the conversion of CO_2 into useful fuels, stability and poor product selectivity of the catalysts. Therefore, synthesis of new nanomaterials with merits of low cost, high efficiency, and outstanding properties is essential. Here, I will mainly discuss recently discovered non-noble metals catalysts, which can potentially replace noble metals for CO_2 electrochemical conversion process.

Nitrogen Fertilizer-Induced N2o Emission Factors for Corn Ethanol Life-Cycle Analysis

Conference: Bioenergy Sustainability Conference

Year: 2019

Proceeding: 2019 Bioenergy Sustainability Conference

Group: Poster Submissions

Session: Poster Session

Authors:

Kwon, H., Argonne National Laboratory
Xu, H., Argonne National Laboratory
Cai, H., Argonne National Laboratory
Xia, Y., University of Illinois
Wander, M., University of Illinois

In existing biofuel regulations such as the Renewable Fuel Standard, benefit of corn ethanol relative to petroleum fuels in terms of greenhouse gas emissions (GHG) is typically evaluated based on life-cycle carbon intensity. It is well recognized that N fertilizer-induced N_2O emissions from corn production contributes significantly to ethanol carbon intensity. Typically, N_2O

emissions from N fertilizers applied are calculated by multiplying N rate with a N_2O emission factor, which could be measured via field experiment or derived from a meta-analytic approach of published literature. Recently, there is a growing interest to capture spatial variations in N_2O emissions and to understand management and environmental drivers.

The objectives of this study are to (1) update fertilizer-induced N_2O emission factors using latest studies and (2) to assess impacts of spatial variability in emission factors on carbon intensity of corn ethanol. Existing N_2O emission database used in GREET life-cycle model was built in 2012 and since, about 280 new papers on this topic were published according to Web of Science. We expand existing database through a meta-analysis of published literature regarding N_2O emissions from corn-based farming systems in US. Literature review covers US-based studies published between 1990 and 2019 to update current N_2O emission factor. Fertilizer management and environmental factors will also be evaluated. Findings obtained from the meta-analysis will be used to characterize the significance of N_2O emissions in the overall GHG emissions profile of corn ethanol production in various US states.

Potentials of Ethanol Development in Africa

Conference: Bioenergy Sustainability Conference

Year: 2019

Proceeding: 2019 Bioenergy Sustainability Conference

Group: Poster Submissions

Session: Poster Session

Authors:

Abeywickrama, S., Sunbird Bioenergy Sierra Leone Limited

Potentials of Ethanol Development in Africa

At a time of record oil prices and rising concern over global warming, ethanol presents a valuable opportunity to reduce dependence on volatile global oil markets, create local economic opportunities in agriculture and industry and build community resilience to climate shocks. In Africa, we are developing ethanol production to develop its huge potentials to create thousands of jobs, generate huge amounts of electricity for the national grid and sequester millions of tons of carbon that could be traded on global carbon markets. Despite the challenges, we see ethanol as a viable product to support Sierra Leone's recovery from the conflict and Ebola epidemic, and to help build sustainable development. As the largest biomass investment in Africa, our poster will present on the current information gaps and the potentials of bioenergy in Africa, based on our experience in Sierra Leone.

Synergistic Effects between Hydrolysis Time and Microporous Structure in Poplar

Conference: Bioenergy Sustainability Conference

Year: 2019

Proceeding: 2019 Bioenergy Sustainability Conference

Group: Poster Submissions

Session: Poster Session

Authors:

Shonnard, D. R., Michigan Technological University
Ankathi, S., Michigan Technological University
Zhou, W., Michigan Technological University
Webber, J. B. W., Lab-Tools Ltd.
Utkarsh, C., Michigan Technological University

Enzyme accessibility plays an important role in controlling the rate of conversion of cellulose to glucose in biomass. The goal of this study is to measure the cellulose accessibility due to the effect of dilute acid pretreatment (DAP) and enzymatic hydrolysis (EH) time of Populus biomass. The NMR cryoporometry technique was applied to measure pore size distribution and pore volume over a wide range of pore sizes (range from 0 to 10 000 Å) for both pretreated and enzymatically hydrolyzed biomass substrates. The results showed a clear shift in the pore size ranges (from smaller to larger) as the DAP (acid concentration 0.5% wt and temperature 160 °C) and EH time increased. In most cases, the pore volume increased with the time of treatment, which clearly suggests internal pore expansion. A linear correlation is observed between the initial rate of enzymatic hydrolysis of pretreated poplar and absolute pore volume greater than the critical pore size (51 Å) and shows the effect of residual enzyme activity on the surface of wood chips.

Techno-Economic Analysis of Biologically Derived Short-Chain Methyl Ketones As a Gasoline Blendstock

Conference: Bioenergy Sustainability Conference

Year: 2019

Proceeding: 2019 Bioenergy Sustainability Conference

Group: Poster Submissions

Session: Poster Session

Authors:

Baral, N., Lawrence Berkeley National Laboratory
Yang, M., Lawrence Berkeley National Laboratory
Scown, C., Lawrence Berkeley National Laboratory

Short-chain (C5-C7) methyl ketones are promising gasoline blendstocks due to high-energy-density and low-hygroscopicity relative to ethanol. These methyl ketones can be biologically produced from plant materials by using *Streptomyces albus*; however, the economic viability of this conversion pathway, key cost drivers, and process optimization and intensification opportunities are not known. To bridge these research gaps, this study quantifies the minimum selling price of short-chain methyl ketones at the current state of the art of the technology and considering potential future scenarios with lignin utilization/valorization. The process model developed in this study includes biomass feedstock supply chain, ionic liquid-based biomass deconstruction method, and the downstream biological conversion, separation, and waste treatment systems. We consider switchgrass as a representative biomass feedstock. We find that the yield of methyl ketones from simple sugars, solid loading rate, sugar yield from biomass, biomass feedstock cost and quality, and coproduct credits are most influential to the selling price of the short-chain methyl ketones. The recently demonstrated methyl ketones yield of 1.6 g/g-sugar results in a very large minimum selling price (MSP) of $163/gge. This selling price could reach to $12/gge with the intermediate product yield improvement of 50% of the theoretical stoichiometric yield. Improving other most influential process parameters along with the product yield of 90% of the theoretical stoichiometric yield reduces the MSP to $2.65/gge, which is just 19% more than the last 10-year average gasoline price at the refinery gate. The results demonstrated in this study could help to develop future research strategies.

Testing Wood-Bark Adhesion for Debarking of Short Rotation Woody Crop: Shrub Willow

Conference: Bioenergy Sustainability Conference

Year: 2019

Proceeding: 2019 Bioenergy Sustainability Conference

Group: Poster Submissions

Session: Poster Session

Authors:

Chahal, A., Penn State University
Ciolkosz, D., Penn State University

Debarking has potential to make short rotation woody crops

economically competitive by both improving the quality of woody material and adding an additional product stream from separated bark. The ultimate strength of wood-bark interface was studied in this research relative to the development and optimization of debarking systems and strategies. Four genetically different varieties of shrub willow, three harvest

times and different moisture levels were tested using universal testing machine, Instron®. The loading rate during tests was 0.5 mm min^{-1} to maintain quasit-static conditions. Regression models showed that moisture content has a negative correlation with wood-bark bond strength, like observations made on other species. The harvesting time of

year also affects wood-bark strength of shrub willow significantly ($p < 0.01$). Shrub willow harvested during the winter has significantly higher wood-bark bond strength compared to spring or summer. This may be due to seasonal changes in cambium layer morphology. Moreover, wood-bark bond strength is significantly different for genetically different varieties of shrub willow. Cultivar â€œMillbrookâ€ has higher wood-bark bond strength than â€œFabiusâ€. It is likely that cross-linkages of calcium ions, and arabinan and galactan side chains of pectins are more common in varieties that have high wood-bark bond strength.

AUTHOR INDEX

AUTHOR INDEX

AIChE
120 Wall Street, FL 23
New York, NY 10005-4020

CURRAN ASSOCIATES INC.
proceedings
.com

ISBN 978-1-7138-0568-7
9 781713 805687